Patterns

David Kirkby

RIGBY
INTERACTIVE
LIBRARY

Cover designed by Herman Adler Design Group.
Designed by The Pinpoint Design Company
Printed in China

00 99 98 97 96
10 9 8 7 6 5 4 3 2 1

Library of Congress Cataloging-in-Publication Data
Kirkby, David, 1943–
     Patterns / David Kirkby.
       p.   cm. — (Mini math)
     Summary: Introduces different types of patterns and
the concept of symmetry with simple activities.
     ISBN 1-57572-003-5
     1. Geometry—Juvenile literature.   2. Symmetry—
Juvenile literature.   3. Tiling (Mathematics)—Juvenile
literature.   [1. Geometry.   2. Symmetry.]
I. Title.   II. Series: Kirkby, David, 1943– Mini math.
QA445.5.K56   1996
516'.15—dc20   95-39647
                                     CIP
                                     AC

Acknowledgments
The publishers would like to thank the following for the
kind loan of equipment and materials used in this book:
Boswells, Oxford; The Early Learning Centre; Lewis',
Oxford; W.H. Smith; N.E.S. Arnold. Special thanks to the
children of St Francis C.E. First School.

The publishers would like to thank the following for
permission to reproduce photographs: J. Allan Cash Ltd,
p. 5; Robert Harding Picture Library, p. 6.
All other photographs: Chris Honeywell, Oxford

contents • contents

You can make patterns by using one or more shapes over and over. These all have patterns on them.

The black and white stripes on a crosswalk make a pattern.

All these things have stripes.
Which things have stripes that
run across them?
Which things have stripes that
go up and down?

• To Do •

Make up your own
patterns with stripes.
Use a ruler to make
straight stripes.
Can you make a curved
stripe pattern?

Stripes make
patterns.
Dots make
patterns, too.

Some animals have stripes.

What makes the pattern on a ladybug?

• To Do •
Copy the dot pattern that makes 6.
Make some different patterns with 6 dots.

Colors can make patterns.
The next fish will be blue.

These beads make a color pattern.

What color patterns can you see?

• To Do •

What pattern do traffic lights make as they change color? Copy the fish on page 8. Make a pattern of your own with 8 fish. They do not have to be in a straight line.

A check pattern has squares of different colors.

This shirt has a check pattern.

10

How many black squares are there?

How many white squares are there?

• To Do •

Design and color
a check pattern
for a flag.

Solid shapes, like cylinders, can make patterns. We call these stacking patterns.

These boxes are stacked in rows. They go across the shelf.

These books are stacked
in columns.
The piles go up.

• To Do •

Stack some blocks of the
same size in rows.
Stack some blocks of the
same size in columns.
Can you stack the blocks
in any other patterns?

Squares tessellate.
The shapes fit
together with
no gaps.

These tiles tessellate.

This ball has 2 different shapes.
Do they tessellate?

• To Do •

Trace around this
shape several times.
Cut out your shapes.
Can you make them
tessellate?

Pictures can make patterns. They do not tessellate.

The dots on the wallpaper make a pattern.

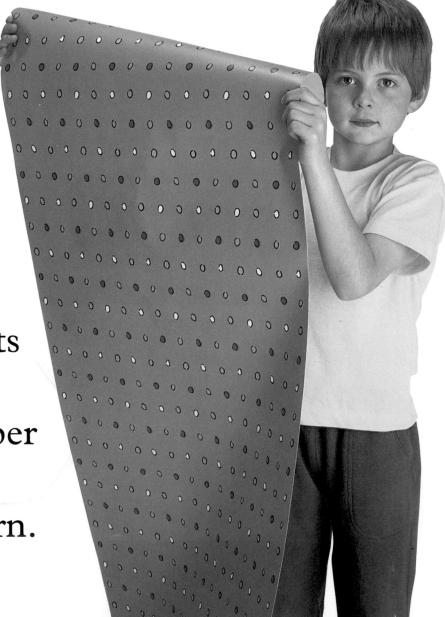

16

Do the stars on the curtains make a pattern?

• To Do •

Draw a rectangle.
Use a ruler.
Make a picture pattern
for a bedspread.

You can find patterns anywhere.

These bricks make patterns.

18

What patterns can you see here?

• To Do •

Copy the sweater pattern.
Design a sweater pattern of your own.

The numbers on
the houses
make a pattern.

The houses on this side of the
street have odd numbers.
The next numbers will be 9 and 11.

The houses on this side have even numbers.

What will the next 2 numbers be?

• **To Do** •

Write all the odd numbers from 1 to 20.

Write all the even numbers from 1 to 20.

Symmetrical things have 2 matching halves.

The 2 matching halves can be anywhere.

Where can you cut this fruit
to make 2 matching halves?

• To Do •
Draw some things
you wear that are
symmetrical.

answers • answers

Page 5	The pillow, T-shirt, and cup have stripes going across. The bag and towel have stripes going up and down.
Page 7	Dots
Page 9	Red, yellow, red, yellow . . .
Page 11	32 black squares, 32 white squares
Page 15	Yes
Page 17	Yes
Page 19	Dots, checks, stripes, picture patterns, diamonds
Page 21	6, 8
Page 23	Anywhere that cuts through the center

24